Icebergs

Other books in the Wonders of the World series include:

Gems
Geysers
Mummies
Quicksand

WONDERS OF THE WORLD

Icebergs

Stuart A. Kallen

KIDHAVEN PRESS™

THOMSON

GALE

San Diego • Detroit • New York • San Francisco • Cleveland
New Haven, Conn. • Waterville, Maine • London • Munich

LIBRARY OF CONGRESS CATALOGING-IN-PUBLICATION DATA

Kallen, Stuart A., 1955–
    Icebergs / by Stuart A. Kallen.
        v. cm. — (Wonders of the world)
Summary: Discusses icebergs, to include: formation; location; color; size;
animals that live on icebergs; and research.
    ISBN 0-7377-1030-6 (hardback : alk. paper)
    1. Icebergs—Juvenile literature. [1. Icebergs.] I. Kallen, Stuart A., 1955– II. Title.
    GB2403.8 .B64 2003
    551.34'2—dc21

                                                                2002009466

Printed in the United States of America

CONTENTS

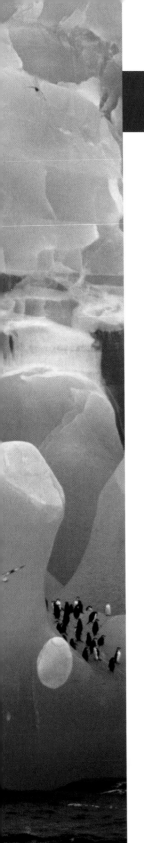

All About Icebergs

For centuries travelers sailing through the icy-cold waters near Greenland and Antarctica have gazed in fear and in awe at **icebergs** rising like giants out of the sea. These chunks of ice are beautiful to behold and may look like towering castles, lofty church spires, or huge marble ships floating along on frigid waters. But long underwater spikes of pointy ice project from icebergs, and these daggers can be as deadly to ships as torpedoes or bombs.

While ocean travelers may fear icebergs, these monster ice cubes are wonders of Earth, as natural as snow and rain. Icebergs are formed from the vast fields of ice that cover the North Pole and the South Pole. When seasonal temperatures warm, icebergs split off, or **calve**, from the main body of ice and fall into the ocean. This is accompanied by a deafening roar and thunderous crash as the ice shatters and falls into the sea with a large splash.

Greenland Iceberg Factories

Icebergs are formed in the two coldest places on Earth— the Arctic around the North Pole, and Antarctica around the South Pole. The icebergs in the Arctic come mainly

Icebergs have many different shapes and sizes. This one has taken the shape of a beautiful floating castle.

As this glacier slowly flows toward the sea icebergs calve and drift away.

from the west coast of Greenland, a large northern island east of Canada that is covered almost completely in ice.

Greenland icebergs calve from icy **glaciers** that are created over the course of thousands of years on the mountains of the North Pole. Snow piles up on these mountains but never melts. As layer after layer of snow builds, it packs down the snow underneath, forming glaciers about two hundred feet below the surface. These glaciers flow outward like rivers of ice, pulled by gravity and pushed by the heavy snow from above. When glaciers flow to the sea, pieces break off and become icebergs. But this takes a very long time—the frozen water that makes up a newborn iceberg may be more than fifteen thousand years old.

The Mertz Glacier Tongue

One of Greenland's largest glaciers is the mile-deep river of ice known as the Mertz Glacier Tongue, which sticks out into the Atlantic Ocean. This formation is an iceberg factory, growing at an incredible pace of sixty-five feet a day and calving up to ten thousand large icebergs every year. Most break off in the warmer months between May and July and float in the northern Atlantic Ocean. Most of these icebergs melt before flowing as far south as the Canadian province of Newfoundland. But each year about fifty icebergs float more than eighteen hundred miles to an area known as the Grand Banks. This is the region where the *Titanic* was destroyed by an iceberg in 1912.

The Mertz Glacier Tongue is only one glacial iceberg factory in western Greenland. Other glaciers produce another thirty thousand large icebergs every year. Smaller icebergs may be as big as a house, while larger icebergs can be several miles across. The average-sized iceberg in this region weighs about four hundred thousand pounds—about the same weight as a fifteen-story building. And although these large chunks of monster ice look beautiful, only about one-eighth of an iceberg floats above the water's surface. The other seven-eighths is underwater.

The tallest iceberg ever seen anywhere was spotted near Greenland. This giant was more than 550 feet high—as tall as the Washington Monument. And this was just a tiny piece of the iceberg. The underwater part went down 3,885 feet below the surface—nearly three-fourths of a mile.

The largest slab of ice ever seen in the Arctic region was recorded near Baffin Island in 1882. This piece of ice was about eight miles long and almost four miles wide. It stuck out of the water more than sixty feet. Its estimated weight was more than 9 billion tons. The iceberg contained so much water that it could have provided everyone in the world a drink of one quart a day for more than four years.

Icebergs of Antarctica

Antarctica has about eight times more ice than the Arctic region. Icebergs in Antarctica are much bigger and are formed differently than those that calve in the Arctic.

Antarctica's landscape is covered with **ice sheets**—large masses of permanent ice that cover both land and water. There, the temperature may dip to -100°F, and snow rarely melts. Instead, it piles up for hundreds of years, creating a sheet of ice with an average thickness of about seven thousand feet. In some places the ice is deeper than the tallest peaks of the Rocky Mountains, almost fifteen thousand feet—or nearly three miles thick.

Overall, Antarctica is one immense ice sheet. It reaches more than 8 million square miles and contains about 70 percent of Earth's freshwater supply. Ninety percent of the world's ice is found here. At the edges of the Antarctic continent, thick sheets of frozen water jut into the ocean in formations known as ice shelves. These

A colony of penguins stands below a huge iceberg in Antarctica.

huge slabs of flat ice make up about 30 percent of the coastline. Scientists have named these formations. The three main ice shelves are called the Ross, the Filchner, and the Ronne.

As the sheets push down into the water, waves and tides push back and huge cracks form, sometimes hundreds of miles long. Eventually, the ice cracks off into icebergs that may be 650 to 980 feet thick and several hundred miles long. These icebergs may weigh billions—or even trillions—of pounds.

The largest iceberg ever recorded anywhere was seen in 1956 near the Ross Sea off Antarctica. This iceberg was more than 208 miles long and 60 miles wide—larger than the country of Belgium.

Icy White, Neon Blue, and Emerald Green

Icebergs the size of a small country can float because they contain microscopic bubbles of oxygen that become trapped between the layers of snow over the centuries. Just as a bubble of air in a closed bottle will allow it to float in a tub of water, the oxygen in icebergs keeps them afloat in the ocean.

These bubbles of oxygen reflect and glitter and give icebergs their brilliant white color. Not all icebergs are sparkly white, however. Those with less oxygen are a striking neon blue color that resemble the polar sky. Icebergs that have absorbed airborne dirt and dust may be layered like cakes with chocolate brown or black ice. But the most astounding—and unusual—icebergs are green.

Travelers in Antarctica have reported seeing icebergs that shimmer a bright emerald color and look like huge scoops of lime ice cream floating in the sea. These icebergs get their color from seawater frozen to the bottom of some ice shelves. This water is laden with microscopic green organisms called **plankton**, a rich source of food for fish and other sea creatures. When the icebergs calve, they take the green ice with them, frozen to their undersides. As they melt, they capsize and reveal their green underbelly above

These icebergs are layered in shades of neon blue and icy white. A cluster of penguins speckles the smooth glassy side of one iceberg (far left).

the surface. While these green icebergs are extremely beautiful, they are also rare. Only one iceberg in a hundred is colored with plankton-laden seawater.

Bergybits and Growlers

Over the centuries sailors have given names to different-sized icebergs. Officially, a chunk of ice must stand at least seventeen feet above sea level to be defined as an iceberg. But smaller chunks of ice have their own names. Floating ice that is shorter than seventeen feet and less than fifty feet long is known as a **growler**, because ocean waves that wash over it sound like the throaty growl of a mad dog.

Bits of icebergs that are smaller than growlers—seven to sixteen feet in diameter—are known as **bergybits**. The tiniest icebergs, those the size of a football or a dinner table—anything less than seven feet in diameter—are called **brash ice**.

These technical names do not come close to describing the many unique shapes of icebergs, however. Descriptions of icebergs are incredibly varied. Some appear to be giant cone-shaped tepees. Animal shapes are also common, including the forms of giant white whales, sleeping dogs, or yawning hippopotamuses. Others see classic shapes such as onion-domed minarets, pinnacled temples, or even King Tut lying in his tomb.

Officially, icebergs that look like fortresses on mountaintops are known as castle or pinnacle icebergs. Long, tablelike icebergs, like those commonly found in Antarctica, are known as tabular icebergs. Others are described

As sailors make their way to shore they must maneuver their dinghy through a maze of bergybits and growlers.

A sheet of ice, broken into bits and pieces, looks like a floating jigsaw puzzle in the sea.

as jagged, blocky, roofed, rounded, and pyramidal, or pyramid-shaped. U-shaped icebergs that are pocked with sparkling ponds of melted water are called dry dock.

Melting into the Ocean

No matter how they are described, wind, waves, and warm temperatures eventually cause icebergs to melt. Some melt in as little as eleven days while others may last more than ten years.

The lower part of this iceberg slowly melts, leaving the upper portion standing on thin, icy stilts.

Icebergs are born deep under the frozen snowpack of Earth's poles. After they calve, they become lumbering giants that rule the sea. And throughout their life, icebergs act as a reminder that the ice and snows of the frigid Arctic and Antarctica are constantly on the move, and forever shifting and changing shape.

Icy Disasters

Icebergs are beautiful works of nature, but they are also deadly to ships. While the small tip of an iceberg thrusting out of the water may be visible to ship captains, the most hazardous part of the iceberg lies beneath the waves. The hidden portions of icebergs are sharp, hard, and extremely dangerous to passing vessels. Boats that hit icebergs have been known to sink in only minutes. Sailors and passengers aboard these ships were doomed when they were forced to dive into the water. In such frigid waters, a person can survive for only a few minutes.

"Iceberg Alley"

Over the centuries, many ocean travelers have drowned in the chilly waters known as the Labrador Current. These currents flow through the shipping lanes in the Grand

Only a small piece of a giant, jagged iceberg shows above the water.

Banks. Every summer they bring hundreds of icebergs south from the Arctic. The most dangerous part of the region is the ship graveyard known as "Iceberg Alley," located about 250 miles southeast of Newfoundland.

The first written record of a ship colliding with a Grand Banks iceberg dates back more than three centuries. On July 10, 1686, the *Happy Return* struck an iceberg. The ship sank immediately, taking the crew down with it.

Back in those days, even those who survived such a collision often suffered. In April 1704, the ship *Anne* hit the underwater tail of a huge iceberg about 150 miles off the coast of Newfoundland. While sailing away with a

The SS *Victoria,* loaded with passengers, crashes into an iceberg off the coast of Alaska in 1880.

large hole in its side, the ship drifted into another submerged ice dagger. As *Anne* sank, its fourteen-man crew was forced to escape in a rickety lifeboat. For seven days these unlucky men drifted in the frigid Atlantic currents until they beached near the town of St. John's, Newfoundland. During the ordeal, five men perished in the boat, and six died soon after landing. Only three members of the crew survived, and one of those lost both legs to frostbite.

Survival in lifeboats sometimes turned ugly after ships collided with icebergs. In April 1841, the *William Brown* hit an iceberg during a storm. Thirty-three of the 83 passengers onboard went down with the ship, while the rest piled into two lifeboats. When one of the lifeboats started to sink because it was carrying too much weight, 17 survivors were thrown overboard to lighten the load. The remaining survivors were later rescued. Only a month later, 120 people were lost when the steamer *President* sank in nearly the same spot.

When the SS *Pacific* sank in 1856, killing all 186 people aboard, notice of the accident was found in a message in a bottle that floated thousands of miles to Hebrides Islands off Scotland. The note, penned by a passenger read "Ship going down. Confusion on board. Icebergs all around us on every side. I know I cannot escape."[1]

The *Titanic* Disaster

In the twentieth century, ships were built larger and more powerful. But icebergs in the Grand Banks area continued to wreak their deadly toll. And as passengers aboard

More than one thousand people died when the *Titanic* hit an iceberg and sank in 1912.

the *Titanic* learned, even a deluxe pleasure cruise can turn into a fatal nightmare after a collision with an iceberg.

On April 10, 1912, the luxurious British ocean liner RMS *Titanic* set sail from Southampton, England, to New York City. This was the ship's first voyage. Aboard were some of the richest and most powerful millionaires in the world. Few worried about the ship sinking. The

The U.S. Coast Guard cutter *Polar Star* patrols the Arctic carrying a crew of scientists.

Titanic was believed to be an engineering miracle, constructed with a double bottom and sixteen separate watertight compartments that ran the entire length of the ship. Engineers believed the ship would stay afloat even if two of the compartments ruptured. The twenty lifeboats could hold only about half of the 2,207 people aboard. The lifeboats were there more for show than for emergency because the boat was considered unsinkable.

Four days into the cruise, on April 14, an iceberg ripped a three-hundred-foot hole in the ship's skin and punctured six of the watertight compartments. Within two and a half hours the *Titanic* was lying on the ocean floor. Only 705 people were rescued. The rest went down with the ship into the icy depths of the Atlantic Ocean.

The International Ice Patrol

The *Titanic* disaster shocked the world and alerted people to the dangers of icebergs. The traveling public demanded that something be done to prevent another such tragedy. About a year after the sinking of the *Titanic*, in November 1913, thirteen countries formed the International Ice Patrol (IIP). This group's job is to track icebergs and warn ships traveling near them.

The IIP monitors the five hundred thousand square-mile area of the North Atlantic. The group tracks more than one thousand icebergs a year. It uses military aircraft equipped with state-of-the-art radar. When conditions are so poor that the planes cannot fly, Coast Guard cutters are used to patrol the area. Far above Earth, satellites also monitor icebergs movements. On land, computers are used to predict the paths icebergs travel.

Bombs and Bullets

The IIP has also experimented with various ways to destroy icebergs that stray into shipping lanes. The ice patrol has used gunfire, mines, torpedoes, depth charges, and even bombs, but with little success.

A massive assault of special heat-producing bombs on one huge iceberg resulted in only a few blackened holes and clouds of black smoke. Hits from torpedoes and shells from five-inch naval guns did nothing. And land mines detonated on the surface provided only a glittering shower of ice cubes.

Researchers think icebergs can be blasted out of the way of passing ships. But the means for doing this carry too much risk. For example, a team of explosive experts

An iceberg is blown up in an attempt to clear the waterway and allow ships to pass through safely.

could land on the ice, drill thousands of holes, and plant dynamite in the holes. But icebergs are unstable, and might roll over at any time, killing anyone on top. Plus it would take hundreds of thousands of pounds of dynamite to damage an iceberg. This would be like lighting 2.4 million gallons of gasoline on fire.

Smaller icebergs can sometimes be towed out of shipping lanes. But IIP has found that the safest and easiest way to deal with icebergs is to simply warn ships away from icebergs, rather than try to move or destroy them.

Another Icy Nightmare

These efforts have paid off. Iceberg accidents have decreased. But in the iceberg-prone North Atlantic accidents sometimes still take place. Forty-seven years after the *Titanic* disaster, the Danish passenger freighter *Hans Hedtoft* repeated *Titanic's* tragic history. Like the *Titanic*, the *Hedtoft* was thought to be unsinkable. It had modern radar, a double steel bottom, an armored bow, and seven watertight compartments. The ship was on its first voyage on January 20, 1959, when it was caught in a storm off the coast of Greenland. Bitter winds and twenty-foot waves battered the ship. When the ship rammed an iceberg, it quickly sank.

By the time rescue planes began to circle the area, all they could see were icebergs bobbing up and down in the churning seas. There was no hope for the fifty-five passengers and forty crewmembers aboard the *Hedtoft*. Their lifeboats would have been instantly swamped in such choppy waters. Officials estimated that no one could have survived more than sixty seconds in the icy cold seas.

Silent Killers

As the *Hedtoft* experience proves, icebergs are a menace to ships no matter how they are built. So when sailors see gleaming chunks of ice thrusting out of the water, they know to go around them. No matter how strong the ship, or how powerful the engines, no vessel can win a battle with a deadly iceberg. While there is great beauty in these natural wonders, they are silent killers of the seas.

Life and Death on an Iceberg

Icebergs can pose great danger to ships. But these enormous floating hunks of ice can also act as homes, protectors, and nurseries for a wide variety of animals—and even humans.

Seals are commonly found sleeping and sunning themselves on tabular icebergs, growlers, and bergybits. These icy beds offer protection to seals, who, while swimming in open water, are easy prey for killer whales. The whales, however, will not swim through fields of icebergs to hunt seals.

Born on an Iceberg

In Antarctica, the seven-hundred-pound Weddell seal spends most of its time underwater, where the sea is a relatively warm 29°F. Few animals, however, experience a

29

A Weddell seal tries to entice her pup into the icy water for a "swimming lesson."

harsher entrance into the world than Weddell seal pups. These pups are born on icebergs, often in howling winds and temperatures that may reach -70°F. The pups come into the world without the thick coats of blubber that protect their mothers. When storms arrive, some mothers dive off the iceberg into the water for protection. Others try to lie near their babies to shield them from the weather. For those whose mothers leave the pups on top of the iceberg nursery, life is short.

Survival aboard the iceberg remains difficult even for those that ride out the storms. When the pups are two weeks old, mothers give them "swimming lessons" by tossing them into the water. After a few minutes, the pup will jump out of the water back onto the iceberg to rest. After a few weeks, the baby seal will finally learn to swim. Then it can join its mother hunting crab in the sea.

Walruses, giant cousins to the seal, also ride the waves on icebergs. These huge animals may grow up to twelve feet in length and weigh up to three thousand pounds. They live in the Arctic and the North Atlantic, where they spend a great deal of time in the frigid waters under the ice searching for clams, shrimp, and mussels to eat.

When female walruses give birth, they use their sharp ivory tusks as hooks to pull themselves up out of the water and onto a passing iceberg. Like seals, they nurse their calves on **ice floes** while teaching them to hunt and swim.

Battling with Bears

Walruses also use their tusks to defend themselves against polar bears, one of nature's most ferocious predators. Polar bears hunt for walruses, seals, and other prey while roaming 12 million square miles of the Arctic ice. Polar bears can also swim up to sixty miles between icebergs. These three-hundred- to five-hundred-pound animals use

A polar bear carefully balances on an iceberg while scanning the water for seals. Tourists take pictures from a boat in the background.

icebergs as hunting stations, floating on the ice while watching the seas below. When seals swim past, polar bears jump into the water and grab them with their huge claws. After catching a seal, a polar bear will drag it back up onto the ice to eat it.

When a polar bear sees a group of walruses sleeping on an iceberg, it will charge out of the water and scatter the herd. While the frightened adults rush into the water, the younger, slow-moving walruses make easy pickings for the bear. Once it catches a walrus, the bear eats it like a banana, peeling back the skin while devouring the contents inside.

Penguins

While huge creatures such as polar bears and walruses wage battles of life and death, little black-and-white penguins face few predators on icebergs.

Adélie penguins of Antarctica find icebergs a perfect place to breed and play.

Millions of penguins live in the cold waters around Antarctica. Although they spend 75 percent of their time swimming underwater, they often breed and hatch babies on ice shelves and icebergs. The Adélie penguin is the most common, and the smallest, of these flightless birds. About 5 million of these creatures live on the Antarctic ice.

When they need to leave the water, Adélie penguins can leap straight up onto icebergs. Where large colonies gather, a parade of penguins may be seen leaping off the icebergs into the sea. These creatures must be careful, however. Their only underwater predator is the leopard seal, which waits quietly beneath the iceberg colonies. If the Adélie does not look before it leaps, it may land in the jaws of a hungry seal.

Life on an Ice Island

Penguins, polar bears, and seals have always lived and died on icebergs. But in recent times, scientists have moved research stations onto large, floating ice islands.

The first iceberg to house humans was named T-3, or Fletcher's Ice Island, named after Joseph Fletcher, a U.S. Air Force colonel who discovered it. This chunk of ice, 9 miles long, 4 miles wide, and 160 feet thick, broke off from the Ward Hunt Ice Shelf along Ellesmere Island in the Canadian Arctic in the late 1940s.

Between 1952 and 1978 the stark-white frozen wasteland of Fletcher's Ice Island was home to a floating scientific research and weather reporting station. The first workers on the island built igloos, carving the pure glacial ice into large cubes for use as building blocks. Large tarps served as roofs.

Arctic Research

Later, a power plant and a runway for aircraft was built. Several insulated huts were flown in to house crews of up to twenty workers. One hut was used as sleeping quarters, while another was filled with kitchen equipment, comfortable furniture, and even a washing machine and clothes drier. A third hut acted as a laboratory, where a weather station and radio were used to broadcast barometric pressure, temperature, humidity, wind speed, and other weather data every six hours.

A research station provides shelter and a good place for scientists to study this isolated environment.

Researchers on the ice island also studied the Arctic Ocean, using equipment to check its depth, salt content, and other factors. Samples of water were tested for plankton and other living creatures.

The iceberg floated for decades at a lazy two miles per hour in a huge twenty-four-hundred-mile circle around the Arctic Ocean near the North Pole. Temperatures of -50°F and whiteout conditions (where blowing snow made it

A team of scientists uses special equipment to measure the depth and salt content of this iceberg.

impossible to see more than several feet) were common. In the winter months the sun does not rise near the Arctic Circle. This continual darkness added to the desolation of the ice island.

Workers moved about on the island's crusty snow with snowshoes and cross-country skis, carrying rifles in case polar bears attacked. Tractors and snowmobiles were used to travel longer distances and move equipment. Food, fuel, machinery, mail, and other supplies were attached to parachutes and dropped onto the island.

The research station, known as a **drift station** operated on Fletcher's Ice Island for twenty-two years, until the iceberg eventually drifted south of Greenland and melted.

Melting Icebergs

Like Fletcher Ice Island, all icebergs melt away in time. But in recent years, scientists have noticed a record number of icebergs calving in Antarctica. And these icebergs are extremely large. High-tech satellites have shown that ice shelves in the Antarctic region are shrinking rapidly. Some scientists think this is a result of **global warming**, a rise in temperatures resulting from pollution from cars, factories, and other sources. Other scientists say the melting of the icecaps is part of a natural cycle of warming that has been taking place for centuries.

Whatever the cause, the melting of Antarctic ice increased dramatically in 1986, when an iceberg covering more than 6,820 square miles broke off from the Larsen Ice Shelf and a similar chunk fell from the Filchner Ice

Some scientists believe that global warming has increased the number of icebergs that calve into the ocean.

Shelf. Around the same time a 990-square-mile piece of the Thwaites iceberg tongue broke off. Together these icebergs were larger than the states of Massachusetts and Connecticut combined—and ten times bigger below the surface. The water from this one iceberg alone would fill half of Lake Michigan with 250 trillion gallons of water or supply the entire water needs of the city of Los Angeles for 500 years.

Since the late 1990s, Antarctic icebergs have been calving at an even faster rate. In March 2002 a major section of the Larsen Ice Shelf, which is 12,000 years old, collapsed into the ocean over a period of 35 days. This iceberg, 733 feet thick and 100 miles by 130 miles, contained more than 720 billion tons of ice. After this latest calving, the Larson Ice Shelf had shrunk to about 40 percent of its average size. Scientists noted that this iceberg was the largest created in the Antarctic Peninsula since the 1970s.

Students take turns sliding down a giant iceberg. There are many ways to enjoy and appreciate these magnificent wonders of the world.

Part of History

Throughout the centuries, icebergs have played an important role in history. They are beautiful natural wonders but also deadly to ships. They act as floating homes to animals and people. When large icebergs calve, they make headlines across the globe. And they reveal to scientists the secrets of the frozen poles. But for all that is known about icebergs, researchers have much more to learn.

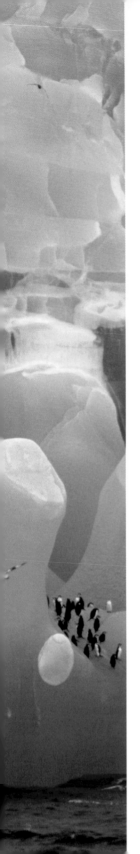

Notes

1. Quoted in Brian T. Hill, "Database of Ship Collisions with Icebergs," National Research Council, May 22, 2001. www.nrc.ca.

Glossary

bergybits: Chunks of floating ice seven to sixteen feet in diameter.

brash ice: Any floating ice less than seven feet in diameter.

calve: To break off from a glacier or an ice sheet.

drift station: A scientific research house built upon an iceberg.

glaciers: Huge masses of ice made from compressed snow that move slowly over land, sometimes falling into the sea as icebergs.

global warming: An increase in Earth's average temperature as a result of human activities such as the burning of fossil fuels (coal, oil, and natural gas).

growler: A floating ice chunk shorter than seventeen feet high and smaller than fifty feet long.

icebergs: Large chunks of ice from a glacier or ice shelf that fall in the ocean and stand at least seventeen feet above the waterline.

ice floes: Flat masses of floating sea ice.

ice sheets: Flat ice formations that extend from icecaps and may cover both land and water.

plankton: Microscopic organisms that float in fresh water or salt water and serve as food for fish and other creatures.

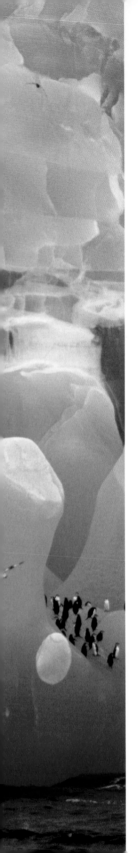

For Further Exploration

Books

Allan Fowler, *Icebergs, Ice Caps, and Glaciers*. Danbury, CT: Childrens Press, 1998. The story of icebergs and their origins.

Arthur McKeown and Peter Hogan, *Titanic*. London: Aladdin Books, 1998. Facts and details about one of the worst sea disasters in history.

Seymour Simon, *Icebergs and Glaciers*. New York: HarperTrophy, 1999. The natural history of icebergs illustrated with many colorful pictures.

Barbara Wilson and Art Wolfe, *Icebergs and Glaciers: Life at the Frozen Edge*. Morristown, NJ: Silver Burdett Press, 1995. The people and animals who live on and around icebergs and glaciers.

Jenny Wood, *Icebergs: Titans of the Ocean*. Milwaukee: Gareth Stevens Children's Books, 1991. Information about floating ice, animals that live on icebergs, and the *Titanic*.

Internet Source

Brian T. Hill, "Database of Ship Collisions with Icebergs," National Research Council, May 22, 2001. www.nrc.ca.

Index

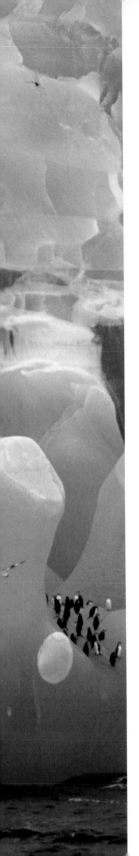

Picture Credits

Cover Photo: © Winifred Wisniewski; Frank Lane Picture Agency/CORBIS

© James L. Amos/CORBIS, 24, 36

© Tom Bean/CORBIS, 38

© Bettmann/CORBIS, 23

© Ralph A. Clevenger/CORBIS, 20

© Corel Corporation, 7, 15, 16, 33, 39

© Dan Guravich/CORBIS, 32

© Hulton-Deutsch Collection/CORBIS, 26

© Wolfgang Kaehler/CORBIS, 8, 11, 17

© PEMCO-Webster & Stevens Collection/CORBIS, 21

© Rick Price/CORBIS, 30

© Robert Weight; Ecoscene/CORBIS, 35

© Winifred Wisniewski; Frank Lane Picture Agency/ CORBIS, 13

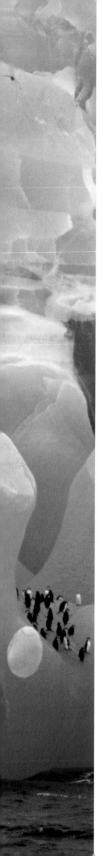

About the Author

Stuart A. Kallen is the author of more than 150 nonfiction books for children and young adults. In addition, Mr. Kallen has written award-winning children's videos and television scripts. In his spare time, Stuart A. Kallen is a singer/songwriter/guitarist in San Diego, California.